The Red Pine
Korean's Tree of Life

Chun, Young Woo

The Red Pine

Korean's Tree of Life

Chun, Young Woo

BooksHill

First published by Book's Hill Publishers, 2009

 For information, address Book's Hill Publishers, 258-20 Suyu-2 Dong, Ganbuk-Gu, Seoul, 142-877, Republic of Korea.

ISBN : 978-89-5526-632-0

The author would like to thank the following authors and publishers for being so generous with their permissions: Prof. Kwang-Il Tak, the Supreme carpenter Eung-Soo Shin, the Asia-Pacific Center for Theoretical Physics, The Korea Foundation, Korea Buddhism Promotion Foundation, and Sumoon Publisher.

Contents

Preface

Pine is pervasive throughout diverse aspects of the Korean culture. Because of the long-term coexistence of Korean people and pines, the utility of the pine, which has sustained the agrarian society across the Korean landscape for the last two millennia, has also transcended into Koreans' world of spirituality and culture. Partial evidence for this spiritual/cultural aspect is that the red pine of Korea (*Pinus densiflora,* or *sonamu* in Korean) has a long continuous history as a subject of literature and artistic pieces. It is also the symbol of life, longevity, integrity, chastity, spirituality, and *poongryo* (festive performance and activity). Currently, most Koreans of the present industrialized society continue their focus to love, cherish, and sustain the red pine of Korea. This long-held admiration continues because the red pine of Korea is the representative icon for the identity of the Korean culture. Understanding of *So-namu* culture (the pine culture of Korea), the unique tree-related culture of Koreans, is more or less parallel with an understanding of the Korean culture, Koreans' views on nature, and Korean forest history.

Over the last twenty years, my curiosity and attention to the role and importance of the red pine of Korea in the Korean culture has been represented in various print media, such as newspaper and magazine articles, and broadcast media, such as TV and radio reports. Some of these thoughts and activities have also been incorporated into a few of my books. One essay (out of the eight articles collected in this book) is a paper coauthored with Dr. Kwang-Il Tak, which was presented at the IUFRO (International Union of Forest Research Organizations) conference held during 2006 in Florence, Italy. Another five articles are based on English translations of articles that were previously published in Korean for various journals and magazines. Translation of these articles was managed by the Asia-Pacific Center for Theoretical Physics (**Crossroads**), The Korea Foundation (**Koreana**), and Korea Buddhism Promotion Foundation (**Buddhism and Culture**). The remaining two articles, which were translated from Korean to English by Dr. Cheong-Ho Yi, were included in the English edition of my book, "Forest and the Korean Culture." Thus, the formation of this book is indebted to the translation efforts by Drs. Kwang-Il Tak and Cheong-Ho Yi and the acknowledged support of the societies and institutions of the publications mentioned above. Also, I extend my gratitude to Dr. Ned Klopfenstein at Forestry Sciences Lab., USDA Forest Service for technical English language editing. I also thank Mr. Seung-sik Cho, the president of Book's Hill publishing company for taking a burden of publishing this book during times of inhospitable economic conditions worldwide.

The purpose of this book is to overcome a relative dearth of English versions of scholarship that represent our Korean academia. Furthermore, my primary motivation for this book is to introduce the non-Korean people to the red pine of Korea, its cultural heritage, and its place within our Korean tradition.

November 2009

Young Woo Chun

Fig▶ The red pine of Korea (*sonamu* in Korean) was the most significant raw material for building palaces (A: *Injeongjeon* Hall, National Treasure No. 225, the main hall of *Changdeokgung* palace first built in 1405), building commercial ships (B: excavated 13th century. *Koryo* ship at National Maritime Museum Korea), and a high-quality energy source for producing pottery (C) and sea salt (D). Parts of pine trees were eaten as famine food (E: a cake flavored with pine endodermis) and used for traditional medicine (F: *Wolfporia cocos* from pine root).

The Red Pine, Korean's Cultural Icon

1

Introduction

Pine trees are essential natural resources to support agrarian society in Korea. Even in the present industrialized world, Korean people still remember that Koreans are born with pine trees, live in pine-wood homes, and are buried in the pine forests behind the village. These fond memories are largely based on the middle and older generations; they started their lives with pine twigs that were fixed in the *geumjul* (sacred straw rope) on their front doors to protect the mother and newborn child from the access by outsiders. Koreans built homes, farming implements, and even coffins with lumber from pine trees. The Korean culture is often called a "pine culture" because of Korean's versatile utility of pine trees.

Role of the Pine Tree in Agrarian Korean Society

No tree has ever been such a driving force to the development of civilization as pine trees in Korea. It is nearly impossible to envision buildings, such as old palaces and temples, without pine trees. The contribution of pine trees to Korean civilization does not stop at buildings. It also helped the transport of people and goods around the country and overseas. Examples of pine ships include the turtle gun-ships that efficiently sunk Japanese's warships and cargo ships that transported

Fig▶ Traditional Korean painting that helps portray the meaning of the saying from the agrarian age, "Koreans are born with pine trees, live in the pinewoods, and are buried in the pine forests behind the village." (*Lee, Ho-shin*, 2005)

rice and salt. The world-famous, white porcelain from the *Joseon* Dynasty was also made possible because of special firewood called '*youngsa*' derived from pine trees. Pine forests especially along the coastal regions helped in the production of sea salt. Among over a thousand tree species that grow in this country, the pine tree provided the most benefits to our Korean ancestors. So precious and essential for agrarian Korea, pine trees were regarded as the "tree of life", and have been managed and protected at national level by special laws called 'Pine Protection Law' for over 500 years. At the community level, each village formed its own '*Songgye*', an organization equivalent to a community forest society or nature conservancy in the western world. The *Songgye* sustainably managed the pine forests close to their village, which was leased from the government on a long-term basis. Presently, some of these forests still remain visible in a few areas. Pine forests in Anmyeondo, which has been protected by the government for over a thousand years, and a *Songgye* in *Bosung*, which has existed for over 200 years, are good examples.

Pine Trees Developed into a Symbol of Korean Culture

The utilitarian virtues of pine also provided an immense influence on the psychological world of our ancestors. The written symbols of universe, life, longevity, integrity, principles, saintliness, and elegance drawn by our ancestors are based on pine trees. Pine trees became naturally integrated into Korean culture, art, religion, ideology, and customs, which subsequently melded and evolved into one symbolic system. This symbol system, which originated from the pine tree and was brilliantly combined with the natural features of Korea's landscape, nourished Korean psyche and emotion.

Other examples still exist today: a charismatic pine tree that symbolizes the minister's rank bestowed by a king, a rich personified pine tree ('*Seoksongnyeong*') that was given legal status as a registered tax payer "who" has paid property tax every year until today, pine forests that were planted when the capital city was moved from *Gaesung* to Seoul 600 years

Fig▸ *Irwoloakdo* was a folding screen to represent the King's dignity and authority. The red pine tree of *Irwoloakdo* represents the symbol of Royal longevity.

Fig▶ *Sipjangsaengdo*. The pine tree, along with sun, mountain, water, stone, cloud, herb of eternal life, crane, turtle, and deer, represented as one of the ten longevity entities. The ten longevity symbols are a manifestation of the human desire to live eternally without aging.

Fig▶ The rich pine tree, '*Seoksongnyeong*' (Natural Monument No. 294), "who" owns farmland and pays property tax like other Korean citizens. In the 1927, a childless farmer couple left their land to the big pine tree in their village. Since then, the tree has been registered as the official owner of the land.

ago, and the reference to the pine trees in Korea's national anthem. All these pine trees and forests maintain a prominent position with modern Koreans. Thanks to the agrarian culture maintained over thousands years, symbols of pine trees are engrained in the Korean psyche as if they are in the genetic code of our DNA. For this reason, the pine trees automatically emerge in our minds, even today, whenever we refer to concepts such as universe, life, longevity, integrity, principles, saintliness, and elegance.

Korea's Pine

Pines or species in the genus *Pinus* are thought to have arisen on this planet about 170 million years ago, and pines represent the the most diverse genus of any conifers. Approximately 110 species of pine are presently recognized, although different authorities accept between 100 and 120 species.

The native pine species inhabiting the land of Korea include Korean red pine(*Pinus densiflora*), Japanese black pine(*Pinus thunbergii*), Korean white pine(*Pinus koraiensis*), Japanese stone pine(*Pinus pumila*), and Japanese white pine(*Pinus parviflora*).

Of these native pine species, the Korean red pine(*sonamu*) is the most numerous and common, with the widest distribution across the Korean landscape. It is often called '*jeoksong*'(赤松) or red pine because the trunk or stem is reddish. It is also called '*yuksong*'(陸松) because it frequently inhabits the interior land, away from the coast. Our native Korean red pine contains two-needled leaves in a bundle or fascicle, so it is considered a two-needled pine. In some regions of its distribution, it grows in a crooked form. However, this tree species also grows with a tree form of straight growth, such as *Keumgangsong* or diamond pine growing in *Gangwon-do* province and the northern part *of Gyeongsangbuk-do* province.

Another native, two-needled pine is the Japanese black pine. Its distribution is mainly in the maritime regions along the eastern, southern, and western coasts of South Korea. It is called '*haesong*'(海松) because it grows in the maritime region. Otherwise it is called '*heuksong*' (黑松) or '*gomsol*' in pure Korean language because its stem color is black. The needle leaves of Japanese black pine are longer and stronger than those of Korean red pine. Japanese black pine is less tolerant to cold than Korean red pine and it occurs primarily in the warmer coastal areas of South Korea.

Korean white pine (*Pinus koraiensis*), Japanese stone pine (*Pinus pumila*), and Japanese white pine (*Pinus parviflora*) are all five-needled pines that contain five-needled leaves in a fascicle. The representative species among these native pines is the Korean white pine. Many plantations or artificial forests of Korean white pine exist in the midland region of the Korean peninsula, and the pine nut of this species is a very delicious, favorite food called '*jat*' in Korea. Japanese white pine is distributed solely on Wooleungdo island, the island in the East Sea. In South Korea, Japanese stone pine is distributed only at the high elevations of Mt. *Seoraksan*, *Gangwon-do* province. Another exotic five-needed pine is the eastern white pine (*Pinus strobus*), which is native to eastern North America. The eastern white pine is frequently planted in plantations at lower and mid elevations, because of its fast growth.

Besides these two- and five-needled species, there are three-needled pines. One of them is '*baeksong*'(白松) (*Pinus bungeana*; lacebark pine), which has been introduced from China hundreds of years ago, is known for its white flaking stem. Other exotic three-needled pines include pitch pine (*Pinus rigida*) and loblolly pine (*Pinus taeda*) which have been introduced from southeastern U.S.A.

Fig▶ The red pine forest regenerated by direct seeding during 1923 to 1926 at *Daekwanryoung* (*Gangwon-do* province, Korea). The red pine of Korea is a native species in Korea that covers 1.6 million ha, one fourth of the total forested land in South Korea.

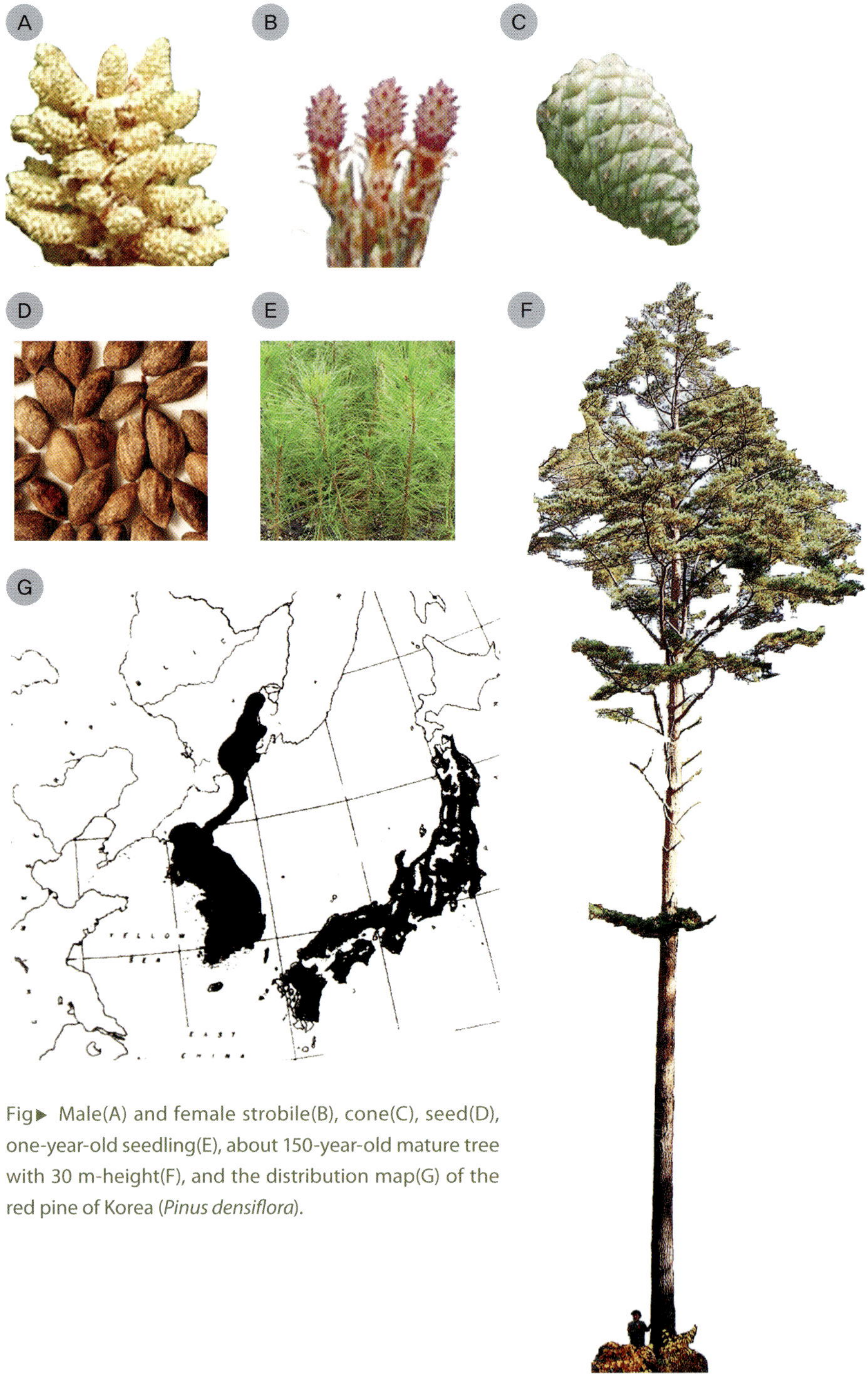

Fig▶ Male(A) and female strobile(B), cone(C), seed(D), one-year-old seedling(E), about 150-year-old mature tree with 30 m-height(F), and the distribution map(G) of the red pine of Korea (*Pinus densiflora*).

Fig▶ The other pines in Korea. A: Native two-needled pine known as Japanese black pine (*Pinus thunbergiii*), B: Native five-needled pine known as Korean white pine (*Pinus koraiensis*), C: Japanese white pine (*Pinus parviflora*), and D: Japanese stone pine (*Pinus pumila*)

Fig▸ *Jeongipumsong* as Natural Monument No. 103 (The red pine of Korea that was bestowed Minister Rank at Mt. *Songnisan*) is a 600-year-old pine tree. Its height is 15 m and its girth is 4.7 m.

The Red Pine Tree, Korean's Most Favorite Tree

2

Trees are magicians that can open the hearts of people. Particularly in my case, as a forestry professor, I cannot help but think that "trees" made it possible for me to have open conversations with strangers. When I reflect on all the people who I have encountered during twenty or so years of my devotion to trees, everyone I met was familiar with the common subject matter of trees, and I have never met a person who disliked trees. However, sometimes issues arise over people's hasty appraisal of my abilities, once I mention my professional background in forestry. Because of this, some people automatically assume that I am some kind of genius who can identify and know everything about each of over a thousand tree species living on this country!

A conversation with a stranger frequently begins awkwardly, but concerns their curiosity about specific trees that have always been around. Some people are curious about why some trees bloom later than others, why other plants cannot bear substantial fruits, and other questions. Once they feel comfortable speaking, conversation flows to various topics about trees; however, at some point, we naturally arrive at the focal point of the conversation, the Korean red pine tree (*Pinus densiflora*).

Under our current situation, various news media remind us daily that pine trees of this country are dying off because of pine-wilt disease.

Fig▶ Fund-raising Exhibition of Pine Paintings by fifteen artists at Seoul to save the pine trees from pine-wilt epidemic, December 2005.

However, upon reflection, I realize that pine trees have always been at the center of conversation about trees and a logical focal point even before the proliferation of the pine-wilt disease became severe. Perhaps this is attributable to the fond memories held by middle and older generations; they started their lives with pine twigs that were fixed in the *geumjul* (taboo cord) on their front doors to protect the mother and newborn child from access by an outsider. This experience contrasts greatly with younger generations, who made their first cries in the maternity wards of hospitals.

The moment the topic of conversation converges on pine trees, we are all happy, our hearts open, and we become closer as we share a common denominator about which we can all care. It does not matter what kind of lives we have led or the kind of work that we do - we all can become engrossed in the topic of pine trees.

A recent experience confirmed my notion that the pine tree serves as the key to opening our hearts. Fifteen artists majoring in Korean

Fig▶ A news conference at Seoul with a hundred people from the culture and art community to urge special government action for the protection of the pine trees from the pine-wilt nematode infestation, March 3, 2005.

Traditional Landscape Art held a fund-raising exhibition to save the pine trees that were dying off because of the pine-wilt nematode (*Bursaphelenchus xylophylus*) epidemic. After the opening ceremony, in the presence of art historians, poets, artists, and others, the main topic of conversation was, of course, the pine tree. Among each participant's discussion of their entangled fates with the pine, the recollections of Professor *Giyun Gwon of Andong* National University were among the most moving. After seeing the old pine trees being cut down on the hill behind his hometown, he said he felt as sad and empty as if he was saying farewell to his ancestors.

From the standpoint of artists who majored in Korean Traditional Landscape Art, it is unquestionable that pine trees, which symbolize traditional values such as principle, integrity, honor, and peace, are an important topic that cannot be ignored. Accordingly, it seems unavoidable that artists drawing Korean Traditional Landscape art would be interested in the pine, which symbolizes the minds and psyche of our ancestors. Similar experiences prompted a hundred people from the Korean culture and art community to prepare a news conference in spring 2005; they urged the government to "legislate a special act for the protection of the pine trees from pine-wilt nematode infestation". Novelists, dancers, painters, poets, publishers, doctors, chief carpenters, ambassadors, professors, and many other artists enthusiastically participated in preserving Korean pine trees, which symbolize our traditional values.

"The National Solidarity to Preserve Pine Trees", which was established during the end of 2005, posed the question of what pine

Fig▸ *Geunjeongjeon* (National Treasure No. 223) is the main hall of the *Gyeongbokgung* palace, which was built of pinewood at the third year of King Taejo (1394).

trees mean to us. This civic group, which possessed a reformative and progressive spirit, established a national solidarity to preserve pine trees, as a physical entity of Korean traditional values. Publicly collaborating to raise their voices and concerns for pine tree preservation might seem unnatural for group of independent artists. However, why did these independent citizens take such organized public action, which could be considered out of character, by establishing a national solidarity to preserve pine trees?

To address that question, I propose to examine the role played by pine trees, which contributed to the farming culture in Korea over millennia. In reality, pine trees were the hidden driving force behind the cultural development of the Korean people. Without pine trees, it would have been unthinkable to construct buildings like royal palaces. When we consider historic times when we were dependent on water transportation, the merit of pine trees becomes even greater. Whether the Turtle Ship or other warships that thwarted the Japanese invaders or ships that transported rice and salt - they were all built by lumber from pine trees. *Joseon* white porcelain pottery, which is world renown and a source of Korean pride, was possible only because of the existence of the pine firewood called "*Yeongsa*". Furthermore, pine forests of the country were responsible for producing salt which was an indispensable basic necessity of earlier times.

In this manner, the pine tree was a vital tree species for maintaining our agrarian culture. A well-established Korean saying, "We were born with pine trees, live among pine trees, and will be buried in the pinery on the rear mountain." illustrates the vital role pine trees play in Korean culture. This saying makes sense when we consider pine twigs laced with *geumjul* (taboo cord) or houses, furniture, wooden boards, and farming implements made out of pine trees. We could name the

Fig▶ Pine trees around at grave yard. Most people believed that even after death, pine trees would bring good luck to the family left behind as well as peaceful rest for the deceased

Korean culture "a pine tree culture" because the diverse utility that pine provides.

Although pine trees provided material benefits for our ancestors throughout the past millennia, this is no longer the case. Today, the legendary utility of pine tree is preserved primarily through the need for pine lumber for cultural restoration projects or through the production of pine mushrooms for gourmet appetites. It is not an exaggeration to say that the role of pine trees as a construction material appears to be almost finished.

Considering that this tree plays a minimal economic role in our current industrialized society, why do we continue to love this tree and fail to forsake it? I believe that this is because the pine tree represents a cultural code for the Korean identity. Pine trees are a symbolic representation of Korean's cultural identity. The pine tree continues to symbolize such values as integrity, honor, devotion, pride, vitality, eternal life, peace, and abundance.

If that is the case, how did the pine tree become a symbol that is so

Fig▶ Two pine trees as the symbol of the Cosmic Tree at the 5th century, *Koguryo* Tomb.

Fig▶ Pine tree as a place of worship. Offerings changed from traditional Korean food and beverages to chocolate, cookies, and even whiskey.

deeply embedded in our mind? We can seek the answer in the lifestyle of our ancestors. Pine trees were outstanding for their diverse utility for construction, shipbuilding, fuel, medicine, as well as relief for famine victims. Among the thousands tree species native to this Korea, pine trees offered the most benefits to our ancestors.

These utilitarian virtues of pines became comingled with its immense influence on the psychological world of our ancestors. The symbols of universe, life, longevity, integrity, principles, saintliness, and elegance drawn by our ancestors are based on pines, and further indicate of the prominent status of pine trees throughout our history. Pine trees became naturally integrated into Korean culture, art, religion, ideology, and customs, which eventually evolved into one symbolic system. This symbolic system, which originated from the pine tree and was brilliantly combined with other natural features of Korea's landscape, nourished Korean psyche and emotion.

The sphere of the pine tree's influence, which melted into Korean literature, art, and symbols of our ancestors, was greater than any other tree species. Pine trees were portrayed as the "tree of life" that appeared

Fig▶ *Wandang's Sehando* (Landscape of Harsh Cold Winter) (National Treasure No. 180). The pine trees of the painting are the symbol of constancy, integrity, and fidelity. (*Kim, Jeonghui, 1844*)

most frequently in classic Korean poems, songs, and landscape paintings. In addition to the beautiful pine trees that inhabit *Irworoakdo* or *Sipjangsaengdo*, the pine tree became infused into folk religion, *feng shui* principles, Confucianism, Buddhism, and Zen philosophy to become a primary element of our national identity. Indeed, the pine tree continues to guide our conscious world to this day.

Regardless of our ancestors' intimate relationship with pine trees in past times, it is difficult to assert that pine trees manifest a current symbolic element of Korean identity unless they become well established in the lifestyles of modern Koreans. A cultural identity that lacks a certain degree of support and popularity by contemporary society would become insignificant over time. Fortunately, pine trees continue to play a role in the cultural identity of Koreans through their personal life experiences, diverse cultural art genres, and various other means.

Fig▶ Pine tree at *Sajikdan* (Altars for the State [Earth and Crop] Deities). The old pine is the symbol of the Royal tree of the *Joseon* dynasty (*Jeong, Seon*, 1676~1759).

Fig▶ Pine trees planted along a community park near a satellite city outside Seoul.

Although the pine trees' role has diminished as a vital resource for agricultural implements, firewood, timber, and hunger-relief food, the pine tree continues to be loved by Korean people as its new roles evolve within the contemporary society. Pine trees are the most popular tree species for planting in landscape projects, parks, green spaces, and borders along historic roadsides. For example, planted pine trees are growing tall in a garden in front of the main office of the Blue House, the Korean president's official residence.

I believe the pine tree has a great influence in unifying Korean society. In our society, which exhibits a typical "mixed dissension", the pine more than suffices as a "supernatural medicine" that can heal the wounds created by generation gaps, disparity in wealth, and regional rivalry. It will not be easy to resolve the multifaceted tensions created through the adverse effects of industrialization and information technology. However, since most Korean people would express their love for pine trees without hesitation, the tree can serve as the source of bonding among diverse people, regardless where they are from, whether they are rich or poor, old or young, female or male, or conservative or liberal.

Although the pine tree no longer performs many of the vital roles it once played during the past agrarian societies in Korea, we must not underestimate the present economic potential of the tree. For example, the total loss from pine-wilt disease infestation of all the pine trees of Korea would cost 19 trillion 200 billion Korean won (19.2 billion US dollars): 12 trillion 500 billion Korean won (12.5 billion US dollars) of non-market and intangible values, 1 trillion 800 billion Korean won (1.8 billion US dollars) of pine mushroom production, and 5 trillion Korean won (5 billion US dollars) of timber value. Furthermore, some estimation suggests that it would cost no less than 16 trillion Koran won (16 billion US dollars) to cut down the 1.6 million hectares of pine forests that have been damaged and to regenerate new forests. What is more important, however, is that it is impossible to accurately calculate economic values for the multiple decades to centuries required for pine trees to grow. For this reason, we can no longer sit idly and continue to permit pine trees to disappear. All citizens must rise and help protect the dying pine trees.

What are the best ways to preserve the pine tree? We can stop or prevent loss of pine trees by active prevention from pine wilt disease, which is known as the greatest current threat. The Korea Forest Service believes that we can prevent the natural disaster spread by an insect called "Japanese pine sawyer"(*Monochamus alternatus*) through appropriate prevention methods; however, an immediate means of reviving or curing diseased trees is currently unavailable. The problem remains not because of a natural proliferation of the disease, rather it continues because of artificial proliferation caused by thoughtless or uninformed people.

Fig▶ Playing the *kayageum* (twelve-stringed Korean zither) in a pine forest.

A recent news report indicated that indiscreet acts of people are spreading diseases with more frequency and speed than the natural infestation. The Japanese pine sawyer can transfer the disease to healthy trees only to the area within a five-kilometer radius, at most. However, people have been illegally smuggling the diseased trees to use the wood as firewood for steam bathrooms, fish boxes, prop wood, wood for landscape works, or other purposes. A critical cause of the outbreak of the disease can be attributed to the shameless acts of people who have spread the disease to places a hundred kilometers away. For this reason, we need more interest and active participation of citizens groups to assure that all Korean citizens are well informed about how to help protect the pine trees.

Another way to help preserve pine trees in Korea is to plant pine trees into the heart of our society. Such approaches can be called "creating cultures with pine trees" or "playing with pine trees". Various events unfolding over the past several years offer several suggestions at the "creating cultures with pine trees" level. Starting from March 2003, the

Fig▶ A dance performance with "joy of the pine" at *Danhosa* Buddhist temple (*Lee, Ae-Joo*, a Human Cultural Asset)

Daily newspaper, *Munhwa Ilbo* has devoted an entire section every week to a column titled "In Search of Pine Trees", which has informed people of the unknown value of pine tree. Through these efforts, a group of pine tree lovers called the "*Solbaram* Society" was established in February 2004.

The "*Solbaram* Society" was formed to discover the true meaning of pine trees, which are firmly embedded in Korean culture; preserve the tree that is gradually disappearing from Korea; and develop and publicize the symbolic and emotional nature of the pine tree for Korean psyche. The society held a news conference with a hundred leading figures from cultural and art communities in Korea and pleaded for our government to "legislate a special act for the prevention of pine-wilt disease for pine trees". In addition, they organized a million-man, signing petition to save pine trees in Korea, as well as to hold a policy forum in the National Assembly to designate the pine tree Korea's national tree. Furthermore,

Fig▶ Reading poems on pine at the pine forest of *Beobheungsa* Buddhist temple (Poet *Park, Hee Jin*, a Fellow of the Korea Art Academy)

along with multifarious cultural artists, they hold painting exhibitions and poetry readings to exhibit their love and care for pine; they conduct various cultural activities related to the pine tree through dance, music, and publishing among others.

We can view the unfolding of these activities an expression of urgency, and we can consider these expressions as a 21st century succession of pine tree culture. Such activities alert people to the value and importance of the pine tree, while mentally and emotionally planting the tree deep into the hearts of our people.

I recollect the poet Park, *Heejin's* praise for a pine tree: "When considering the surprising vitality it shows to persevere through the hundreds of years it took to grow from a seedling to an ancient tree, and when considering its perfect self realization, the profound harmony between an individual tree and a pine forest grove, the unique status of the pine tree, and its elegance among others, the pine tree is very well the symbol of the spirit of our people. It is more than fair to say it is a great model."

Fig► Pine forest at Misiroyung pass(*Gangwon-do province*)

Pine Forests, the Face of the Land of Korea Altered by the Agrarian Culture

3

Although many other tree species occur across the land of Korea, why is Korean red pine the one that most often comes to mind first? Is it because this pine is the tree that is most commonly distributed across the land? Over the last 60 years, the pine forest in Korea has been reduced to a third of its previous extent, due to the massive exploitation by Japanese colonialists, illegal logging, and overcutting in times of social disorders like the Korean War and other unstable periods of South Korean societies that occurred after the country's Independence in 1945. However, 23 percent of the entire forest area in South Korea is still occupied by Korean red pine forest.

The Korean red pine's position as the favorite tree of Korean people is likely not because it is the most numerous and widely distributed across in the land. So, why does the Korean red pine occupy such prominent position in the mind of today's Koreans? The prominent position of Korean red pine is more likely because this tree is the distinguished product of the agrarian culture of Korea, which has been fostered by our Korean ancestors for millennia. Throughout their history, Korean people have experienced pain, pleasure, and all other human emotions in close association with the pine trees.

The evidence that Korean red pine is the product of agrarian culture

Fig▶ Human intervention in pine forests over the several centuries has altered natural forest succession toward an artificial climax condition and subsequently maintained pure pine stands near villages as an 'artificial climactic forest'.

is apparent even today. For example, you can find vibrant pine forests near some villages where people lead their lives today, while this tree cannot be easily found as you go deeper into the mountains where human settlements are absent. In other words, the pine forests appear to have formed naturally, but they are not as they appear. In contrast, these forests have retained their shape and structure largely because of continuous human intervention. Professor G. Jahn at Goettingen University, Germany, described this aspect of Korean pine forests with the term "artificial climactic forest".

Today, the pine forest in Korea is gradually declining and disappearing as human intervention has been discontinued. An excellent example is present on Mt. Namsan in the center of the metropolitan city of Seoul. Forested areas where the pine forests once thrived are making a transition

Fig▶ The reduction of rural populations due to industrialization and the shift to fossil fuel as an energy source has eliminated the need for human interventions into the pine forests. This human intervention previously maintained pine trees as an artificial climactic forest .

to oaks, such as *Quercus mongolica*, as a result of natural processes of forest succession in the absence of human intervention.

Thus, the reduction in rural population and energy conversion from fuel-wood to fossil fuel, a result of modern industrialization in Korea, decreased human intervention in pine forests near villages. The pine forests, which had been previously maintained at the artificial climax stage, are now succumbing to the natural ecological forces of forest succession and slowly being transformed into broadleaved forests.

Our Pine Grows Well Even in Infertile Soil

How has the Korean red pine become the predominant tree across the land in Korea? Furthermore, why is the distribution and abundance of this critically important tree currently in a state of precipitous decline? Before our human ancestors began to settle and live across the land of Korea, the original composition and structure of the forests must have comprised broadleaved trees. These prehistoric forests existed in a stable condition, composed of deciduous broadleaved trees, such as like oaks and maples. However, these forest began to change as more people populated the land of Korea. The shift toward pine trees arose because our human ancestors developed their civilization by using forests in the land of Korea, similar to what occurred in other civilizations across the world.

Based on the rice seeds that were unearthed in the peat layer of an archeological site at *Ilsan*, *Kyunggi-do* province, it is known that rice cultivation was already being practiced 4,000 years ago in the plains of the southwestern Korean peninsula. In addition, it is reported that forest destruction to extend land available for cultivation was occurring 2,500 years ago in the midland region of the Korean peninsula. Based on this evidence, it can be inferred that the agrarian culture had matured in the society more than 2,500 years ago, at a time when our Korean ancestors had begun to tend and smelt iron at a large scale.

After a long period of the nomadic lifestyle associated with the hunter-gatherer society, the pressing societal need was to solve the problem of maintaining the fertility of cultivated land that would allow the establishment of settlements in selected places. Human excretion, ashes from a fuel-hole or *ondol* (a heating system for Korean traditional housing), and manure made of grass bunches and leaves of broadleaved trees, which were collected from village neighborhood and put to the process of putrefaction, were used as a means to retain the soil fertility of cultivated fields. However, the locally available excretion from men and livestock, ashes from burned organic materials, and other amendments

Fig▶ The pine forest, which could form even in the places where broadleaved trees cannot form forests, is not so different from a natural climax forest.

were not enough to maintain soil fertility at a satisfactory level. To further augment soil fertility, the human settlers gathered the grasses and leaves of neighboring mountains of low elevation and made them into the manure. The leaves were especially collected from broadleaved tree forests in the neighboring lands and mountains. The collection of wood from broadleaved trees, which were cut for the fuel-wood used in an *ondol* during winter, and the collection organic matter from the forest floor of the neighboring forest ushered in a great change to the structure of the forest used by the people.

Thick, broadleaved tree forests surrounded the original human settlements and villages in Korea. However, people continuously cut the broadleaved trees to provide fuel-wood for heating and collected

the leaves of the broadleaved trees to produce the manure needed for agriculture. As a result, soil fertility began to decline in the places near human settlements and villages. Subsequently, the neighboring forests near the villages also became further degraded so that the broadleaved trees could no longer grow. Only pines, which could grow under conditions of low soil fertility, could take root and grow. In this manner, pine trees began to gradually replace broadleaved trees in forests that

Fig▶ Natural broadleaved forests at the National Arboretum, *Gwangreung, Kyunggi-do* province.

grew near to villages.

This scenario is supported by results from pollen analysis, which show that pine started dominate the Korean forests about 1,400 years ago. This period corresponds to the transition time from the Three Kingdom Period (BC 57~AD 654) to the Unified *Silla-Balhae* Period (AD 654~935), for which considerable evidence is found from the agrarian culture and the iron culture that had greatly expanded and advanced.

The human behaviors, such as retaining soil fertility and utilizing the heating system of *ondol* to protect against winter cold, are associated with settlement in fixed places and a shift away from the hunter-gatherer society's nomadic lifestyle. These human behaviors played a crucial role toward altering the structure of the forests adjacent to human settlements, and gradually helped to convert the original broadleaved forest to a pine forest.

All Trees Except Our Pine Are Miscellaneous

Once forests near human dwellings were converted into pine forests, what caused them to be retained as pure pine forests? Though several other factors can be inferred, continuous human intervention appears to be the primary reason that pine trees retained their predominance in the Korean forest. In proportion to population growth in the *Joseon* dynasty, which was continued from the *Koryo* period, the uses and value of pine wood increased and extended.

As forests in the populated regions were transitioning into pine forests, the governmental regulations on trees and forests began to form gradually. Historical records and documents demonstrate that these forest/tree regulations were gradually established and became more formalized as the royal or national institutions at the beginning of the *Joseon* dynasty (AD 1392~1910).

During the initial period of this dynasty, *Joseon* had no logical alternative but to put strong forest-protection policies into practice. Because of administrative disorder at the end of the *Koryo* dynasty (AD 918~1392) before *Joseon*, forest destruction had become so severe that Taejo, the first king of *Joseon*, set out to protect forests by proclaiming the prohibition of private ownership of mountains and forest at the beginning of the *Joseon* dynasty.

Wood from Korean red pine was used not only as core building materials for building the palaces and houses, but also for the material to construct ships, the most important means of transportation at that time. Because of this recognized need for pine trees, the government of the *Joseon* dynasty wanted to protect pine forests for the purpose of maintaining the supply of wood.

Thus, the government of the *Joseon* dynasty recognized the need for pine protection as the representative forest policy. This policy can be summarized in the Korean phrases, '*songmok geumbeol*'(松木禁伐), which means forbidding the cutting of pine wood, or 'songgeum'(松禁), which

Fig▶ *Whangjang* prohibition landmark (*geumpyo*) at Gurongsa Buddhist temple in Mt. *Chiak*, *Wongju*, *Gangwon-do* province.

means pine-cutting is forbidden. In an abbreviated form, this forest policy regarded all the other trees as "miscellaneous trees" or 'japmok'(雜木). As a result of this policy, the common people of the *Joseon* dynasty could freely use the broadleaved trees that were regarded as "miscellaneous". As a consequence, the human intervention (mainly caused by the common people) worked to continuously select against the broadleaved trees growing in the pine forests by direct cutting or use of the organic matter from the forest floor. These activities helped to continuously maintain the pine forests as pure stands.

Since that time, however, the pure pine forests have declined due to the extensive periods of exploitative cutting. Appropriate measures that

allowed and promoted the growth of pine trees for a few centuries were not in place. In the pine forests of some populated regions, the trees with commercially valuable straight forms were continuously cut out, while the pines with crooked forms were left behind. This continual process of removing the straight trees, gradually converted the pine forests so that they comprised badly shaped and slow growing trees, which contributed to eventual degradation and decline of these pine forests. The most readily

Fig▸ A new forest is being formed by seeds from a mature tree at *Sogwang-ri* in *Taebaek* mountain range.

apparent pine forests of Korea, which the modern Koreans would recognize, are the result of processes that Koreans have been practicing for thousands of years in the agrarian culture by sacrificing the broadleaved trees in the forests.

It is Time for the Human Touch to be Applied Again

Because of the crooked shape of its trunk, the Korean red pine was once regarded as more appropriate for landscaping purposes rather than construction material for diverse uses. Some people regard it as a "weak" tree that is susceptible to diseases and insects, such as the pine-needle-gall midge, and frequently has to be cut down. Others regard the Korean red pine as the tree that, if untended, gives up its inhabiting space to oaks. In this vein, Korean red pine was considered as the tree of declining nation, a misnomer based on false ideas. Is the Korean red pine really a weak and useless tree?

Nobody can find a tree to take the place of Korean red pine in the land of Korea. Korean red pine grows vigorously in the climate of Korea, withstanding the dry spells that occur in every season except

Fig▶ Pine forest at *Daekwanryoung* (*Gangwon-do* province, Korea). The red pine of Korea is a native species in Korea that covers 1.6 million ha, one fourth of the total forested land in South Korea.

summer and the continental climate with large fluctuations between the temperature highs and lows. No other tree except Korean red pine can be identified for the land of Korea, which was devastated through the exploitation by Japanese colonial rule and unstable Korean societies after the Independence in 1945. Korean red pine may be the best tree species for the forest soil of Korea, because two thirds of the entire forest area of Korea has soils composed of granite or gneiss, which cannot be easily restored into the original state.

What can we, the Koreans, do for Korean red pine, which incorporates the national spirit of Korea and has become part of the Korean psyche? To stop the disappearance of Korean red pine forests from the land of Korea, all Koreans must pursue the modern human touch or intervention of behalf of the pine forest. Such modern intervention can replace the human interventions put forth during the agrarian culture period. The modern human touch must develop forest nurture work and provide ample investment for tending Korean red pine forest in a manner based on sound science.

Fig▸ *Geumgangsong* (a Korean word meaning Diamond pine), the prototype of the red pine of Korea at *Sogwang-ri* in *Taebaek* mountain range.

Preserving the Original Korean Pine Forest of *Uljin's Sogwang-ri* Village

4

The pine is Korea's most common tree, and *Sogwang-ri* is an area where the original pine trees are best preserved. Hence, it would be proper to describe the forest there as Korea's representative forest. In spite of today's extensive roadway network, the *Sogwang-ri* forest is not so easy to visit, for it is nestled in a valley in the far north of *Gyeongsangbuk-do* province, surrounded by the *Taebaek* mountain range.

Until just a few years ago, the road leading into *Sogwang-ri* was narrow, unpaved, and difficult for passenger cars to traverse. The road is still unpaved but it has been widened and compacted to accommodate vehicles. It would be easy to rush right past the valley running along both sides of the *Sogwangcheon* stream, the source of *Buryeong* (Buddha Reflection) valley. It would indeed be a pity to miss the enjoyment provided by this area of such picturesque scenery.

The pines surrounding the valley do not differ significantly from those within it. However, I would recommend that you stop here for a moment to take in the marvelous landscape. If you are blessed with the opportunity to visit this valley during May or October rather than the crowed months of summer, you can park alongside the roadway and enjoy the trees stretching out as if to embrace the valley. Spring's freshness, visible in the glistening reflection of the crystal clear rivulets cascading over large, flat boulders, or the radiant forest basking in the autumn glow, reveals a truly unique natural beauty. Gazing at the streams reflecting the deep blue sky and pines, one realizes that the realm of the divine hermits, often mentioned in traditional Korean stories, is not necessarily so illusory after all.

The only village in the *Sogwang-ri* valley is located about 4 kilometers from National Road 36. Those who have not arranged for food and lodging elsewhere can stop by the village store, which doubles as a home stay. Slash-and-burn farmers used to live scattered about the valley. However, after armed communist guerrillas infiltrated the *Samcheok-Uljin* area in October 1968, everyone had to reside in one area, which became the village.

At *Sogwang-ri* hidden away in this remote mountain valley, you can see, in addition to the splendid forests, many cultural relics related to the pine tree. The pines near the ancient forest-preserve marker, the shamanic altar, and the traditional Korean house built from the wood of a single

Fig▸ Traditional Korean house built of the wood of a single pine tree at *Sogwang-ri.*

tree are good examples. The latter catches your eye immediately upon entering the village. Built of timber from a single tree, this house vividly demonstrates the traditional relationship that the village has maintained with pine trees. Providing enough timber to build an entire house, the original tree must have been truly magnificent.

Fig▸ The sacred tree with a taboo cord wound around its trunk is the village's tutelary deity.

Another site worthy of attention is the village's sacred tree, which has long been associated with shamanic traditions and beliefs.

What makes this tree noteworthy is the light it can shed on the Korean people's philosophy toward nature. Rising skyward as if to pierce the heavens, it stands some 300 meters to the north of the village. Situated on a hill to the left of where the *Sogwangcheon* Stream merges with the *Taegwangcheon* flowing in from the right, this sacred tree with a taboo cord wound around its trunk is the village's tutelary deity. One should pause before this tree for a moment to appreciate its majestic splendor

A *hwangjangbongpyo*, also known as *hwangjanggeumpyo*, is an inscription made by order of the royal court on a rock face which states that the adjacent forest, due to its exceptionally high-quality timber, was to be protected and nurtured, while being designated as off-limits to unauthorized cutting. During the *Joseon* Dynasty (1392~1910), whenever there was a death in the royal family, fine tan-colored pinewood was used to produce double coffins in which royal family members were buried. This kind of pine was called *hwangjangmok*.

Hwangjangbongsan were protected as royal forest preserves in order to ensure a ready supply of *hwangjangmok* for the royal coffin craftsmen known as *jaegung*. Mountain *hwangjangmok* preserves were called *hwangjanggat* or *hwangjangsan* in Korean. The latter term referred to mountains with stone inscriptions designating the areas as off-limits to the public to protect *hwangjangmok* pines that had already reached maturity. The *hwnagjangbongsan* system was initiated in 1680, the sixth year of the reign of King *Sukjong*.

Fig▶ Warning against the illegal felling of trees, stone markers were used during the *Joseon* Dynasty to protect forests of high-quality pine trees reserved for use by the royal household.

Another worthwhile place to stop on the way to the pine forest is the area next to Sogwangcheon Valley, where the forest preserve marker of *hwangjangbongpyo* is located. This is the most recently discovered *hwangjangbongpyo*. Three others had previously been found with locations at the entrance to *Guryongsa* Temple on Mt. *Chiaksan* in Wonju, *Keunjeorgol* at *Hangye-ri, Inje*, and *Hwangjanggol* Valley at *Dusan 2-ri, Sujumyeon, Yeongwol*, all are in *Gangwon-do* province.

To reach the *hwangjangbongpyo*, follow the road from *Sogwang-ri* toward the *Taegwangcheon* stream. After crossing four bridges and then going up the road some 200 meters, you will find the inscription on the mountain slope between the bottom of the valley and the road. This *hwangjangbongpyo* has stood here for centuries, practically unnoticed, in silent witness to history. The inscription is carved onto a natural rock with a single line of four characters on the left and five lines with 19 characters on the right.

The four characters on the left are 8 centimeters in height and 3 millimeters in depth, reading "*Sanjik Myeonggil*"(The mountain keeper's name is *Myeonggil*). The four rows of 19 characters on the right read: "The border of this hwangjangbongsan extends from *Saengdal* Hill to Mt. *Anilwangsan* to the *Daeridangseong* Fortress."

Sogwang-ri is therefore

Fig▶ The fine tan-colored pinewood was called *Whangjangmok* (above), and it was used as coffins for royal family members of *Joseon* dynasty (below).

the only location where a visitor can see, in one place, a traditional *hanok* (Korea wooden house) built from a single tree, a sacred pine tree with the village's shamanic spirit altar, and an inscription marking a forest reserve whose trees were used for making coffins for the royal family. However, our main interest is the pine forest, which serves as a poignant symbol of the unique Korea character. If you walk about 3 kilometers along the forest road alongside the *Sogwangcheon* stream up from the *hwangjanggeumpyo*, you will observe groves of *geumgangsong*, the original Korean pine. I recommend taking the national forest road that cuts through *Satgatjae* Ridge at 1,119 meters elevation on its way to *Seokpo*. This location allows for a breathtaking view of the pines, especially during early or late summer, before or after the summer

Fig▶ A 500-year-old red pine tree standing at the foot of the trail to *Seokpo*.

vacation season. As a fire-prevention measure, this area is off-limits to the public from late autumn until spring, when the first buds appear.

At the entrance to the forest road where it heads sharply upward toward *Satgatjae* Ridge, a 500-year-old pine rises majestically with a commanding presence. This point is a great place to should stop and get out of your car. To fully appreciate the essence of the *Taebaek* mountain range pines, one should cross Satgatjae Ridge on foot, if possible.

The *Yangbaek* region (northern *Gyeongsangbuk-do* province) situated between the *Taebaek* and *Sobaek* mountain rages has been revered from ancient times as an area of especially high-quality pines. For two decades from the mid-1950's to the mid-1970's, the local *Chunyang* Train Station was well known as a collection and transport point for the area's timber. The pinewood transported from here to places across the peninsula was known to consumers as *Chunyang*mok or *Chunyang* timber. Even today, the pines from this area are more often referred to by this name rather than *geumgansong* pines.

In the past, anyone passing through the *Buryeong* Valley area between *Bonghwa* and *Uljin* would have been undoubtedly awed by the straight *geumgangsong* pines reaching toward the sky. These extraordinary pines, with trunks over 30 meters tall, are the source of dense, tan-colored wood that was unrivaled as a high-quality building material from ancient times.

The *geumgangsong* pines, nestled between the *Yangbaek* mountain ranges, were famous even beyond Korea's shores. The *Koryuji* Temple in Kyoto, Japan, is home to a famed wooden sculpture of the Avalokitesvara Bodhisttva (Goddess of Mercy) sitting in a crossed-legged contemplation pose. This statue, which is considered to be the grandest artistic masterpiece of Japan's Asuka Era (ca, 7th century), was designated National Treasure No. 1 by the Japanese government. Indeed, this masterful work is said to be comparable to Rodin's "The Thinker" and the ancient Greek's "Venus de Milo". However, resembling the gilt-bronze Maitreya Bodhisattva image (National Treasure No. 83) in a similar posture that was produced during Korea's three Kingdoms period, the statue is believed to have been created in the *Silla* Kingdom and then transported to Japan. This hypothesis is based on the fact that a significant number of gilt-bronze and stone contemplative figures (those sitting with one leg crossed over the other) have been found around *Bonghwa, Gyeongsangbuk-do*. Thus, it can be asserted that many such figures were also created from wood of that area, and gives credence to further consideration of this Korean origin hypothesis.

If this hypothesis is true, it would be reasonable to assume that the wood used to create the Japanese sculpture was from the *geumgangsong* pine forest. Even though this contemplative figure is the treasure of another nation, it certainly enhances our appreciation of Korean pines when we consider how our ancestor's creativity went into fashioning this wood into something that became an invaluable cultural heirloom for the entire world.

Pinewood from the *Taebaek* mountain range was thus used to create a leading national treasure of another country. Yet when we look at these pines, a perplexing question arises that cannot be easily dismissed. The trees of the *Taebaek* range grew straight upward for more than 30 meters. But, why do the pines of more densely populated areas of central, western, and southern Korea have curved or crooked trunks? This question is even more intriguing when you consider the pine forests growing around *Gyeongju* and the nearby areas of *Cheongsong, Bonghwa*, and *Uljin*. Compared to the pines with straight trunks and tapered crowns that are growing in the *Cheongsong* and *Chunyang* areas, the *Gyeongju* pines have crooked trunks and slightly flared crowns. The explanation for these

Fig▶ The smiling Maitreya Bodhisatva thinker statue, Japanese National Treasure No. 1 in Koryuji Buddhist temple in Japan (left) and the bronze smiling Maitreya Bodhisatva thinker statue, Korean National Treasure No. 83 excavated in *Gyeongju* (right).

different growth forms can be found in the way in which the *Silla* people utilized their forests.

For more than 1,000 years, the *Silla* people in the *Gyeongju* area selected only high-quality pines with straight trunks from the forests around them to use as fire-wood and construction lumber for their homes and palaces. Because the straightest and best trees were constantly being cut down, the trees of inferior quality were left to propagate offspring. This process was perpetuated for centuries. After over 1,000 years of this selective process, the only trees that currently remain in the *Gyeongju* area are crooked and unsightly.

Those who research cultural history related to forestry often point to Roman's penchant for hot baths as a major cause of its downfall. To assuage the public's growing discontent with the political situation, the Roman authorities continually provided hot water to the baths free of charge. As a result, the forest areas around Rome and other distant locations within the sprawling empire were cut down for firewood to heat the baths. This indiscriminate destruction of forest areas to maintain the hot baths resulted in the degradation of soil quality and vital farmlands, which caused a decline in production and eventually led to the fall of the Roman Empire.

From my perspective as a forestry scholar, I will speculate that suggest that *Silla* may have followed a path similar to Rome, by wantonly destroying its forests and exhausting its resources to the point of ultimate collapse. In *Gyeongju*, the *Silla* capital, over 40,000 tile-roofed houses existed at the time of the kingdom's peak. According to ancient records of that period, a royal edict required that residents use only charcoal for cooking and heating, so as to save pine branches and fire wood for other purposes.

The impact of using such vast quantities of firewood to bake roof tiles for houses and produce charcoal for over 40,000 households, as in the case of the Roman baths, would have inevitably led *Silla* down a destructive and irreversible course. It thus seems reasonable to speculate that over time, the forests around *Gyeongju* became barren, which led to the degradation of the area's arable land and the decline of the nation's fortunes. In support of this theory, scholars believe that the flowering of *Silla*'s Buddhist culture relied on the availability of extensive forests. Numerous *Silla* monasteries created large temple bells, such as the Divine

Fig▶ Majestic pines were prevalent when the *Silla* civilization was in full bloom, but the pine form deteriorated over time.

Bell of King *Seongdeok*, or *Emille* Bell as it is popularly known (National Treasure No. 29). To cast such a bell, copper must have been smelted from copper pyrites, requiring temperatures of over 1,000°C for this process. In this manner, tremendous quantities of firewood were consumed from surrounding forests.

The pine forests in the *Gyeongju* vicinity, which fueled the creation and maintenance of such a vibrant culture, eventually fell into ruin. As a result, even the pines around the royal tombs, which would have been generally preserved, are now, like those around them, bent and twisted. As such, it is not surprising that the pines of the *Taebaek* mountain range - an area not far from *Gyeongju*, but highly inaccessible prior to the 1950's - remain high-quality specimens even today.

It is a pity that the *Sogwang-ri* forest no longer enjoys the fame it once held. According to historical stories, trees, an arms-span wide, were felled and even after being cut into three of four pieces, ox carts still struggled to transport the logs. However, such massive trees are no longer found in this area. It is thus remarkable to hear stories passed on by local elders, who will tell you: "The trunk was so wide that after cutting the tree down, six people were able to sit on the stump and have lunch!"

The pines of this area did not escape the plunder of Korean forests during the Japanese colonial occupation, and likewise, during the era of the *Syngman Rhee* government. Then in the 1970s, these pines were further victimized by the pine-needle-gall midge, an insect that attacks pine trees.

The good news is that our government, recognizing the value of the area's pines, is continuing efforts to preserve and manage the forest by designating it as a protected area. We must all do our part to protect this natural treasure. Although it will take decades or centuries to regain its former majesty, if nurtured and preserved over time, there is no reason that it should not develop into a magnificent forest of unsurpassed splendor. With this in mind, last year (2000) the Korea Forest Service, along with the Citizen's Movement to Grow the Forests of Life, designated the *Sogwang-ri* forest "a beautiful forest that must be preserved as we move toward the 22nd century".

Fig▸ The *Gumgangsong*(Diamond pine) road near the *Shinsyesa* temple

Diamond Pine Forest at *Shingyesa*, Mt. *Geumgangsan*, North Korea

5

- Fulfilling the Duty of the Guardian Deities

A piece of rice cake and a bottle of water - it may seem simple. However, this buddha's birthday meal that was offered to us in May 2007 touched our party deeply. Earlier in the day, our group had hiked on Mt. *Geumgang* through the rain, and went to *Guryong-yeon*, one of the three major waterfalls in Korea. By the time we stopped by *Shingyesa* Temple on our way back down the mountain, we had been totally soaked by the spring rain. To us, the temple's celebration of Buddha's birthday is unforgettable. It seems like a dream that we were able to celebrate this day in North Korea with a South Korean monk.

To take cover from the rain, we all crowded under the eaves of the temple's Great Hero Hall, where a monk greeted us warmly. *Jejeong Sunim*, a South Korean monk, was the planning director of the *Shingyesa* Restoration Project. I told him that my students and I were visiting as part of a Forestry Cooperation Seminar between South and North Korea, and we came to observe the tree nursery and chestnut tree plantation in Mt. *Geumgangsan*. Then, I asked him to give us an overview of the restoration project.

Jejeong Sunim began by mentioning that *Shingyesa* was one of the four major temples of Mt. *Geumgangsan*, which were all destroyed during the Korean War. Now *Shingyesa* is being rebuilt through the joint efforts of the Buddhist communities in both Koreas. He further believed that this effort would help open the way for their non-violent reunification and provide a place to worship the Buddha of Reunification. He added that this breakthrough Buddhist project was only possible after certain auspicious clauses and conditions were met: The South Korean would provide wood and technology; the North Koreans would provide water, sand, earth, and historical research.

Hard work on the rebuilding project progressed for 10 years since *Gosan Sunim* held a commemorative restoration ceremony in 1998. In 2001, a first-ever geographical survey was conducted around the *Shingyesa* temple grounds at the time of the Lantern Festival. In 2002, a memorandum of understanding (MOU) was signed between the two Koreas, and in 2004, the ground breaking ceremony was held. *Jejeong Sunim* was appointed as the planning director to supervise the project.

The restoration project began with the completion of the Great

Fig▸ *Jejeong Sunim*, a South Korean monk and students of Kookmin Univ. at *Shingyesa* temple

Hero Hall (main Buddha hall) before proceeding with restoration work on the five remaining buildings. After ten years of hard work, the completion ceremony was held on October 13, 2007, attended by *Jigwan Sunim*, the President of the *Jogye* Order of South Korea, and *Yoo Young-sun*, chairman of North Korea's Buddhist Federation. This practically concluded the duties of *Jejeong Sunim* as the planning director of the restoration project, and he subsequently returned to South Korea.

After wrapping up his comments on the temple rebuilding, the discussion naturally moved on to the beautiful pine forest standing tall, like guardian deities, around the temple. The view of the *Shingyesa* stream in front of the temple's Great Hero Hall was breathtaking, with rain mists covering the peaks of the surrounding mountains and the lush green forests creating a sublime panorama.

Fig▶ The Restoration Completion ceremony of rebuilt *Shingyesa* temple on October 13, 2007.

Fig▸ *Shingyesa* temple and pine forest with mountain peaks in the distance.

It was our good fortune that no other visitors were present in *Shingyesa* temple, except our group of twenty people. In addition, *Hyundai Asan* provided a bus for us to visit the sapling nursery, which had been established with funds from a variety of South Korean civic groups. Free from crowds, we were able to comfortably view the beautiful scenery around *Shingyesa* accompanied by just *Jejeong Sunim*.

Jejeong Sunim talked with much enthusiasm about each aspect of the pine forest surrounding the *Shingye* Stream. Because I had previously experienced the profound majesty of the Diamond Pines as a pine lover, I was already strolling through the trees in my mind while listening to his words. Perhaps he felt that his praise of pine trees had run too long, and he started asking questions to the students. "Where are you from?"

Fig▶ The red pine forest at Mt. *Munphil-bong* near *Shingyesa* temple

We answered, “We're from *Kookmin* University,” to which he made an unexpected reply: “I am familiar with Prof. *Chun, Young-Woo* at *Kookmin* Univ. through the book he wrote.” My students quickly informed him that I was the author, making me feel sheepish. Unaware of those around us, the monk and I shared in a very involved and lengthy discussion on pine trees.

It is interesting to note that the pine trees around *Shingyesa* temple are widely renowned. First of all, their unique value is attested to by the North Korean government's designation of this pine forest as Natural Monument No. 416. It's easy to see why the forest is protected as a Natural Heritage site, as the pine trees here are straight and dignified, giving rise to their ancient nickname, "Pine Belles". The pine trees around *Shingyesa* are alternately called "Diamond Pines", reflecting their immense value. Some areas of *Gangwon-do* Province and North *Gyeongsang-do* Province also have pine trees that are referred to as "Diamond Pines", a name that

Fig▶ The pine forest surrounding *Shingyesa* Buddhist temple has been designated as Natural Monument No. 416 by the North Korean government.

most likely originated from the towering pines of Mt. *Geumgangsan* (diamond).

About 80 years ago, a Japanese forestry scientist, Prof. *Uegi Homiki* conducted a nationwide survey of the pine trees in Korea. Upright pine trees that were distributed throughout *Gangwon-do* Province, North *Gyeongsang-do* Province, the *Taebaek* Mountain Range, and *Yeongdong* (the middle-eastern part of Korea) were classified as 'Diamond Pines', and Prof. *Uegi Homiki*'s appellation has remained a part of our forestry terminology ever since.

My first brush with the pines of *Shingyesa* temple began in 1999. At the onset of Mt. *Geumgangsan* tourism in North Korea, many restrictions were applied. I recall that our group was given only 20 minutes to look around the *Shingyesa* temple site. The weary tourists returning from *Guryong-yeon* were not overly anxious to get off the bus again, as we had been following a tight schedule. However, this stop was a highlight for me. I immediately darted out of the bus and made my way around the pine forest surrounding a three-story stone pagoda, thoroughly enjoying the majesty of the straight Diamond Pines around me.

When I visited Mt. *Geumgangsan* for the

Fig▸ The *Shingyesa* temple's three-story, stone pagoda and the pine forest before the temple's restoration (1999)

Fig▶ Withered pine trees suffering from an infestation of the pine-needle-gall midge (August 2006)

second time in 2006, my mission was to evaluate the progress of the chestnut and pine tree nursery. Thanks to the generosity of the staff there, I was given my first-ever opportunity to hike to my heart's content in the area around *Shingyesa* temple. However, I was surprised to see many bare, dried-out pine trees suffering from predation by the pine-needle-gall midge, which had flown in from the direction of *Wonsan* City. I was full of worry when I returned from my hike.

When I visited the forest for the third time in 2007, we were overjoyed to see that the health of the pine trees around the temple had recovered. Unfortunately, because of the recent deterioration in South-North relations, I don't know when we will be able to again see *Shingyesa* temple and its surrounding pine forest. I remain hopeful that the two Koreas will realize a "bloodless reunification" and that the Buddha of Peaceful Reunification enshrined at the temple will one day allow us to visit the North once again.

Fig▶ A ceremony of harvesting pine trees for restoring *Sungnyemun* (National Treasure No. 1, South Gate of Seoul) which was burnt down February, 2008.

Gyeongbokgung Palace, the Main *Joseon* Palace, under Renovation with Korean Red Pine

6

According to the characteristics that distinguish one period from another, Korean culture can be divided into three periods: the stone culture of the *Silla* kingdom (BC 57~AD 935), the clay culture of the *Koryo* dynasty(AD 918~1392), and the wood culture of the *Joseon* dynasty (AD 1392~1910). Remnants of the *Silla* kingdom include diverse stone artistry, such as *Seokgulam*, the *Seokga* pagoda, and the *Dabo* pagoda. *Koryo* has left the legacy of beautiful celadon pottery, of which, the color did not change even after a thousand years. *Joseon* contributed the wooden furniture that is unadorned and simple, but is so distinguished in its form and style. As Korean ancestors have made their living for several thousand years in the land of Korea, they left the diverse legacy of artworks made of stone, clay, and wood produced from the very land they inhabited. This legacy can be readily observed in the National Treasures and Valuable Treasures of Korea.

National Treasures are the most precious treasures in a country. Among the valuable entities of cultural legacy of a country, the buildings, artworks, and craftworks of the highest academic and artistic values are selected and designated as National Treasures. Korea has about 260 National Treasures and 20 of them are related to wood. The wooden National Treasures include the 14 wooden buildings like *Sungraemun* the

Fig▸ *Cheonmadojangni* (National Treasure No. 207, Saddle flap with 'heavenly horse' painting) made of birch bark.

Great South Gate to the walls of the capital, and *Kyunghoeroo* building in *Gyeongbokgung* palace and the main shrines or *daewoongjeon* of some old Buddhist temples. In addition, our National Treasures also include six artworks made of wood of various tree species.

Cheomando jangni (Korean National Treasure No. 207) unearthed from the *Cheonma chong* or the tomb of heavenly horse is the picture depicting a heavenly horse on the birch-bark plate for a saddle flap. The Tripitaka plates at the Haein Buddhist temple (Korean National Treasure No. 32) and the *Koryo* Buddhist canon plates (Korean National Treasure No. 206) are the wooden plates made of Japanese cherry, stone pear, and birch, which are inscribed with texts of Buddhist canons written in ancient Chinese characters. *Mokjo samjonbulgam* (Korean National Treasure No. 42) at the Songgwangsa Buddhist temple or the *mokjo munsoodongja* statue (Korean National Treasure No. 102) are Buddhist statutes made of wood (*mokjo*) of unknown species. *Hahoe* masks and *Byungsan* masks (Korean National Treasure No. 102) are wooden masks made of alder wood.

Among about a thousand Valuable Treasures of Korea, 95 entities are related to wood. Most of these 95 Valuable Treasures are buildings constructed of wood. The fact that there are so many National Treasures and Valuable Treasures made of wood indicates that wood was an essential natural resource that was indispensable for our Korean ancestors.

The Wood Culture of Our Korean Ancestors

Throughout the long history of Korea, our ancestors continually developed and managed their own technology for utilizing diverse wood materials. Due to the distinct topography of mountainous terrains occupying most of the land of Korea, our ancestors developed technology to utilize diverse tree species that were readily acquired from the places close to their dwellings. Houses were built with pinewood, which could be obtained from nearby mountains, and home furniture and agricultural tools were also manufactured of various woods. The cooking and heating energy was derived from burning the twigs and branches from miscellaneous trees or *japmok* (mostly broad-leaved trees).

The conditions under which our Korean ancestor lived did not necessarily favor a culture based on wood. The tools and instruments for wood craft were developed relatively late and a circulatory network for logwood distribution was largely inactive due to the inconvenient overland transportation. Despite these obstacles, our Korean ancestors created a culture based on wood that was fully dependent on the properties of trees growing close to their dwellings to provide diverse applications to benefit their own lives.

The landscape characteristics of Korea, which is considerably mountainous with only interspersed plains, led our Korean ancestors to develop a distinct culture of dwelling. Whenever possible, they constructed buildings in harmony with their surrounding natural environment. Our Korean ancestors believed that high, multi-story buildings and lofty mountains are all of 'yang' (陽), which conflicted with each other. They further believed that if these two *yangs* (high buildings and mountains) were to meet, the land vitality would decline accordingly, a result inferred from principles of the yin-yang concept and Five Element theory. For this reason, they established wooden buildings of "proper" height rather than those of loftiness; they also produced furniture and domestic tools that matched well with the size and form of their wooden buildings. Our ancestors' concepts and views of nature are based on the development of this wood culture. They wanted to lead a life in harmony with nature and therefore, negated the possibility of massive construction projects or other remodeling of nature that transforms nature according to human will.

The Korean red pine, which is native to the land of Korea, played a critical role in shaping and consolidating the wood culture of Koreans. Wood from the Korean red pine has some disadvantages as construction material: the wood cleaves relatively easily compared to that from other exotic species; straight long logs are not so numerous; and the wood sometimes undergoes twisting, even after the logging process. However, these limitations as construction material have been complemented and overcome by the dexterous workmanship and remarkable construction technology of our Korean ancestors.

The traditional Korean construction technology is sometimes considered to be relatively coarse in its processing touch, while reflecting a lower precision. However, these traits were derived from the acclimatized, wood-utilizing wisdom of our Korean ancestors, who endeavored to make most of the wood properties of pine wood. The results of such acclimatized, wood-utilizing wisdom can also be observed

Fig▶ *Hungryemun* (the 2nd gate of *Gyeongbokgung* palace), which was renovated with red pine wood.

in *Gyeongbokgung* palace today.

Gyeongbokgung Palace is being Renovated Using Materials from Our Korean Red Pine

Gyeongbokgung palace is the main palace of the *Joseon* dynasty. Since its establishment by *Seonggye Yi*, *Taejo* of *Joseon* about 600 years ago (1392), the palace has been vulnerable to damage and by fire, especially during the Korean-Japanese War (1592~1599) or Japanese colonial rule (1910~1945). However, the palace has been maintained in remarkable condition, due to continued restoration efforts.

In 1990, a major restoration project was initiated for the symbolic legacy of the *Joseon* dynasty, *Gyeongbokgung* palace, which suffered damage or total destruction of some portions during wars and other events after colonization. As the first outcomes, *Gangnyong-jeon* (the King's bedroom-living room suite), *Gyotae-jeon* (the Queen's bedroom-living room suite), and *Jaseon-dang* (the Crown Prince's bedroom-living room suite), *Bihyon-gak* (the Crown Prince's annex), *Heungryemun* (the second gate into the palace) were restored to their original shape and structure. Subsequently, the areas of *Taewon-jeon* and *Gwanghwa-mun* (the main gate to the palace) will be restored. Sometime in the near future, after restoration is completed for all the basic components of the *Joseon* palace, everybody will be able to see *Gyeongbokgung* palace in its original form. The restored *Gyeongbokgung* palace complex with include the complete palatial system that comprises *jeongjeon* (courtroom suite), *pyonjeon* (King's suite), *binjeon* (Queen's suite), and *donggoong* (Crown Prince's suite).

Many people listen to news reports about progress in the restoration of the *Gyeongbokgung* palace. They learn restoration progress, such as the deconstruction of the governor-general's and that *Gangnyong-jeon* and *Gyotae-jeon* will be restored anew. However, most people who follow the restoration progress of the *Gyeongbokgung* palace, the symbolic legacy of *Joseon* dynasty, are unaware the this restoration is based largely on the precious wood of the native Korean red pine. The *Joseon* dynasty used only Korean red pine wood for constructing the palaces. Thus, the same pine wood is also being used for the restoration project of *Gyeongbokgung* palace, so that this restoration will match the original construction. However, why was this palace constructed of pine wood in the first place?

To get detailed information about why pine wood was used in palace construction during the *Joseon* dynasty, I met the supreme craftsman or supreme carpenter for the restoration effort, Mr. *Eung-Soo Shin*. Fortunately, Mr. *Eung-Soo Shin* painstakingly spared some time for me during a busy time, and we met right at the restoration site of

Fig▶ The supreme artisan, *Shin, Eung-Soo*, a carpenter member of the Human Cultural Assets, stands in front of the standing renovation site of the *Gyeongbokgung* palace in the capital of *Joseon*

Gyeongbokgung palace. The supreme carpenter or '*dopyonsu*' is the leading craftsman at the highest rank in the traditional construction. Another highest-rank carpenter in the field is called '*dodaemok*'. But *dopyonsu* is the chief of the royal craftsmen or palace craftsmen called '*pyonsu*', while *dodaemok* is the chief of the leading carpenters (*daemok or daemoksu*) in general, who build the big-house buildings, including Buddhist temples.

To build a great number of big buildings in a palace that symbolizes the authority and power of a despotic king or emperor, 12 kinds of craftsmen or engineers were required to accomplish this grand civil engineering. The craftmen and engineers included wood-craftsmen or '*mokjang*'(木匠) who tend wooden materials, stone-craftsmen or '*seokjang*'(石匠) who tend stone and rock materials, roof tile-craftsmen or '*gaejang*'(蓋匠) who treat roof tiles and construct the roof, and brick craftsmen or '*jeonjang*'(塼匠) who work with bricks, and '*dochaejang*'(塗彩匠) who work with coloring processes or '*dochaejang*'(塗彩匠) who paint and give colors to all kinds of surfaces. To construct a palace of such grand magnitude, it is necessary to create an efficient division of labor among the various types of craftsmen and engineers who possess high-level workmanship and engineering skills. However, such construction efforts also require thoroughly consistent collaboration and cooperation among the diverse types of craftsmen and engineers. Therefore, to maintain a thoroughly consistent operation of a palatial construction project, the supreme craftsman (*dopyonsu*) had command over the twelve chief craftsmen (*pyonsu*), who were the leaders of the twelve kinds of craftsmanship. The craftsman roles must be recreated for the major renovation projects.

For the construction of elite-class houses and Buddhist temples, the wood of *Zelkova*, oak, and fir was also used along with that of Korean

Fig▸ *Gyotaejeon* (Queen's bedroom-living suite) of *Gyeongbokgung* palace renovated in 1994 with the wood of the native red pine of Korea (*Pinus densiflora*).

red pine. However, only pine wood was used for the construction of the palaces where the royal family lived. Mr. *Eung Soo Shin*'s provided a simple answer as to why only pine wood was used for palace construction. It was because Korean red pine was considered as the supreme tree, because it was the most commonly found and possessed the highest strength properties among all the native trees of Korea. Additionally, he told me an interesting episode that conveys the great wood values of Korean red pine. This incident concerns the wood strengths of the pine trees from Korea and those from the current Chinese territory side of Mt. Changbaisan. He imported the pine wood from Mt. Changbaisan since he could not acquire the thick-girder wood required for the restoration work of *Gangnyong-jeon* and *Gyotae-jeon*.

Fig▶ Beauty pine or Changbai pine (*Pinus sylvestris var. sylvetrisformis*) at Mt. Changbaisan, China.

Before it was processed to make the refined wood material for construction, the logwood from the South Korea and Mt. Changbaisan, China was piled up in a designated open site for a few years. After this time, the pine wood from South Korea remained of high quality, without any defects. In contrast, while the imported pinewood from Mt.

Fig▶ Hinoki cypress (*Chamaecyparis obtusa*) at Akasawa Recreational Forest, Giso, Japan.

Changbaisan had become so degraded by rain and wind that it could no longer be used for the thick-girder wood. The wood strength acquired through adaptation to the climate and environment in Korea over million of years represents the true character of the native Korean red pine tree, which is unparalleled by any other tree species in Korea.

I recalled that Hinoki cypress (*Chamaecyparis obtusa*) had been used for the main building (*geum-dang*) of Horyuji (Horyu Buddhist temple) in Japan, which is the oldest wooden building in the world. I asked Mr. *Eung Soo Shin* about the wood quality of Hinoki cypress. The *dopyonsu*'s answer was resolute, without hesitation. He said that Hinoki cypress provides the best wood for building construction in Japan, but the wood of Korean red pine is the best for building contruction in Korea.

The *dopyonsu*'s trust in the pine wood was quite solid. The 500- to 600-year-old, traditional buildings, which were constructed with the wood of Korean red pine, were scrutinized in detail. Even today, the

Fig▶ Trimmed pine timber for renovation of *Gyeongbok* palace.

wood materials still retain their strength properties and other desired characteristics without much degradation. It is believed that these wood materials should be able to sustain the buildings for at least a few hundred more years. These exemplify why there is no other wood better than that of Korean red pine for construction. This *dopyonsu* firmly believes that many wooden buildings of more than one thousand years in age would still exist Korea, if not for the foreign invasions, wars, and social disorders that devastated the wooden buildings so severely. This firmly held belief contrasts greatly with the false, commonly held idea that Korean red pine is a useless tree species, which only occurs with a crooked form and ugly shape. In truth, Koreans have no need to be envious of Japan for its Hinoki cypress.

Mr. *Eung-Soo Shin*, the supreme craftsman, added that it was not a deterioration of the wood strength of Korean red pine that has caused collapse of the many old, wooden buildings in Korea. The wood strength itself remains high over time. Instead, degradation or total collapse of the old, wooden buildings of Korea is attributed to bad maintenance and clumsy renovation applied to wooden buildings or inferior roof-tile assembly.

I further confronted Mr. *Eung-Soo Shin* with my question about whether the pillars of pine wood, which have relatively many knots, can give enough wood strength for sustaining the heavy, tiled roof of a traditional building. Again, his answer was positive. If the pine wood contains live knots rather than dead knots, the knots in the pine wood themselves protect against cleavage in horizontal direction. So, the knots in pine wood do not necessarily diminish its value for construction.

High-Quality Pinewood from *Taebaek* Mountain Range

One of the main tasks that confront the supreme craftsman is securing the pine wood needed for the building materials. As an example from a real restoration project, 2 million pieces of pine logwood (an amount of about 500 trucks carrying 11 tons of freight) was used only for the restoration of the *Ganyoungjeon* (King's bedroom-living room suite) and *Gyotae-jeon* (Queen's bedroom-living room suite). When the thick-pillar wood was unavailable for palatial construction, Douglas-fir from North America could sometimes be used in place of Korean red pine. However, the pines either from New Zealand or Chile were not a suitable replacement, because they are softer and lack the necessary wood strength.

While explaining about the situations where foreign pine wood is unavoidably used, Mr. Shin spoke an acrimonious criticism toward Korean foresters who make their living based on forests. Efforts to keep and manage Korean red pine forests must not be stopped for any reason. If such conservation efforts were forsaken and appropriate forest management in the field was curtailed, the lack of Korean red pine wood

Fig▶ The red pine forest at *Taebaek* mountain range

will cause an absolute tragedy. Historical restoration projects in the future cannot be successfully accomplished while relying solely on imported wood. I was extremely humbled by Mr. Shin's soliloquy.

Last spring, to obtain information for a magazine article, I had an opportunity to visit the forest of Hinoki cypress in Giso, Japan. Most impressive was the attitude of the Japanese foresters, who maintain detailed records of logging and forest management activities. The Japanese have been conserving the Hinoki cypress logwood sites for renovation of Yeesei(伊勢) Shinto Palace (jinku, 神宮) as their forest cultural legacy in Giso. After returning home from Giso, my first activity was not writing the magazine article, instead, I gave a phone call to Mr. Shin, the supreme craftsman. I wanted to ask him about the availability of the logging records for the woods used in the restoration of *Gyeongbokgung* palace.

Our question and answer session continued later at the restoration

Fig▶ The harvest of red pine at the *Daekwanreung* area for renovation of Seoul *Soongryemun* (South Gate of Seoul), which burned down in February, 2008.

site for *Gyeongbokgung* palace. The supreme craftsman concurred that *Yangyang* at the foot of Mt. *Seorak* was the source of the best pine wood. It is in this region that, Korean red pine (*Pinus densiflora*) with big red heartwood is produced. As a reminder, the Korean red pine's common forestry name is derived from its bark color of red namely *jeoksong*(赤松), and not the color of its heartwood. However, the supreme craftsman re-classifies the Korean red pine into three types, depending on the heartwood color and growing district. The first type is red-heartwood pine that grows in the coastal regions of *Yongdong* district, the eastern part of the *Gangwon-do* province, exposed to the maritime wind. The second type is an inland pine that grows in the *Youngseo* district, the western inland part of the *Gangwon-do* province. The third type is a very rare white-heartwood pine, which possesses heartwood that has become white after so many years of being red.

According to the supreme craftsman's explanation, red-heartwood pine is stronger in wood quality than inland pine, in terms of wood strength, cleavage, twisting, and crooking. Red-heartwood pine growing in the maritime wind has relatively greater resin content than inland pine grown in harsher conditions, such as infertile or rocky soil. Its heartwood is red and the ratio of sapwood is meager compared to the heartwood. Inland pine grows two times faster than red-heartwood pine. However, the inland pines' heartwood is yellowish and the wood strength is weaker than red-heartwood pine. For this reason, logwood for the restoration project of *Gyeongbokgung* palace is selected from red-heartwood pine grown in regions such as *Yangyang* (*Samsan-ri*, *Myunokchi-ri*, and *Jang-ri* at *Hyunbuk myun*), *Myungju* (*Sagimak-ri* at *Sacheon-myun* and *Eoheul-ri* at *Seongsan-myun*), and *Samcheok* (*Icheon-ri at Wondeok-eup*).

Fig▸ Trucks loaded with red-pine logs at *Daekwanreung* Forest.

One tantalizing fact is the unavailability of a systematic report that summarizes the production districts

Fig▸ *Gumgangsong* (known as Diamond pine, a form of the red pine in Korea) filled with golden yellowish heartwood.

of Korean red pine wood used for the restoration of *Gyeongbokgung* palace. While rummaging through his documents and records, the supreme craftsman declared that the pinewood production districts can be readily examined. Furthermore, photographs of each district and the associated logging operation could be found without much difficulty. However, he was curious about why I was trying to find this information.

It is indeed unfortunate that our Korean ancestors did not maintain suitable documents and records about the trees and forests. If the records from a few hundred years ago were still available, the shape and structure of our Korean forest would likely be considerably different than what is observed today. If suitable forest records were available, we would know the appropriate tree species that are well adapted for planting and growth in each district, based on its use and growth across the land of Korea for hundreds of years. Future generations can also rely on historical forest to manage and grow healthy forests in the future. The future generations of Koreans provide a strong justification as to why the current production districts of the Korean red pine wood for *Gyeongbokgung* palace restoration must maintain suitable records. If this information is recorded, future generation will know the production districts (known as provenances in forestry) where superior pinewood is found or grown. Without this information, our future generations will have to start all over from the very beginning to determine information that we already know. For this reason, appropriate forest records reduce the time and effort needed by our future generations to maintain and restore the precious cultural legacy of Korea.

Fig▸ The red pine forest at *Anmyeon-do* recreational forest, *Chungcheongnam-do* province.

The Role of Pine Forests in Shaping Korean Traditional Cultural Landscapes

7

Introduction

Pine forests in South Korea have diminished by 40% over the last 40 years because of factors such as climate change, insect and disease damage, human-mediated species replacement, ecological succession, and forest fires. Today, pine forests have been reduced to only 23.5% of the total forested area in South Korea. This rapid decrease of pine forests has caused the decline in traditional landscapes that had previously existed in various forms, such as windbreaks, riparian forests, and scenic forests around villages. One promising sign among the many negative changes is that more pine trees are being planted along the streets, in parks, and for landscaping works. Another encouraging sign is that the importance of traditional cultural landscape is being recognized. Government and citizens groups have started joint efforts through various campaigns to protect pine trees and forests to maintain our living cultural heritage.

Pine Forests as Traditional Korean Cultural Landscape

Most Koreans readily recognize pine trees as a symbol of the Korean mind or as an integral part of traditional culture. This symbolic meaning of pine trees has been shared, accumulated, developed, and continually evolving over time to become a cultural code representing the Korean identity. Some clear examples of pine forests representing Korean cultural identity can be found in the forests established for *feng shui* principles and security. Pine forests for protecting tombs of dead souls, and other forests of the same pine are located around villages for protecting the well-being of living people have contributed to the representative cultural landscape of Korea.

Pine Forests around Tombs and Cemeteries

Being evergreen with plush green needles, pine trees have been regarded as a symbol of everlasting, non-surrendering, and non-declining nature. Therefore, those who aspired to a long-lasting destiny and family prosperity planted pine trees around their ancestral tombs, and those trees became part of typical traditional cultural landscape in Korea. Such roles of pine trees as cultural landscape elements still continue to this day.

The pine trees planted around the royal tomb and the common people's tombs represent the belief that pine trees will bring a good luck to descendants.

Pine Forests around Capital Cities

Pine trees have long been planted around national capitals to fulfill *feng shui* principles. The *Koryo* dynasty (AD 918~1392) founded the nation's new capital in Songak. The history of the *Koryo* dynasty records a story that shows how a new nation could be successfully founded by planting pine trees around the denuded, rocky mountain in *Songak*. The *Joseon* dynasty government chronicle records that the *Joseon* dynasty, which followed the *Koryo* dynasty, also planted pine trees in Mt. *Namsan* at the center of the dynasty's capital, which is now Seoul. The reason that these two dynasties planted pine trees around the capital was based on our ancestors' belief

Fig▶ The pine trees planted around the royal tomb and the common people's tomb represent the belief that pine tree will bring a good luck to descendants.

Fig▶ Old pine trees lining a road of Seoul in the 1920's

that pine trees symbolize the longevity and prosperity of a dynasty.

Pine Forests around Palaces

Pine trees, a tree of life, were planted around Royal residences for *feng shui* purposes. Pine trees appear in '*Dongkwoldo*', a panoramic painting showing the King's residential buildings of *changdukkung* palace and *changgkyungkung* palace. Large individual pine trees and several pine groves were established across the palace complexes during their construction and occupation, which is in stark contrast to the lack of trees today. Currently, no sign exists of such large pine trees and groves; however, photographs from the late 19th and the early 20th centuries clearly show that large, old pine trees were within those two palaces. Scenic images of the landscape, created by old pine trees around *Okryucheon Soyochung*, were captured by American astronomic scientist's photography in 1883. Tall and straight pine trees around *Inchungcheon* and *Daechocheon* also appeared in the book '*Corea Coreani*' by Carlo

Fig▶ *Donggwoldo* (National Treasure No. 249). A panoramic painting of the *Changdeokgung* and *Changgyeonggung* palaces portrays many old pine trees.

Rosetti, who was the Italian Consular General in Seoul from 1902 to 1903. These records support the fact that pine trees played a critical role in shaping Korean traditional cultural landscape around palace.

Pine Forests around Scholar's Residences

The pine tree was a tree most beloved by Korean scholars, who almost always wanted to have pine trees around. In the history of three ancient Korean Nations, *Choi, Chi-won*, a well-known scholar at the time, is recorded to have built gazebos and summer houses, planted pine trees and bamboo. After having traveled through forests and along the rivers and seashores of the Korea, he spent the remainder of his life reading hundreds of books. Pine trees were the most-preferred tree species of the *Koryo* people, who commonly planted pine and Korean pine (*Pinus koraiensis*) in their backyards. Scholars also planted pine trees around their residential homes and Confucian schools to mimic and practice the tree's ever green and unyielding vitality, fidelity, and integrity, which all symbolize the scholarly virtues. Today, pine trees are still standing

Fig▶ Pine trees standing around the *Seongyojang* (traditional scholar's house) at *Kangneung, Gangwon-do* province, Korea.

around old traditional buildings and Confucian schools to demonstrate their role in the traditional cultural landscape.

Pine Forests around Villages for Security

Pine forests were also established around villages to protect from natural disasters, such as storms, salt spray, and flooding. The forests established for this purpose are called 'village forests' and also played an important role in shaping the Korean traditional landscapes across farming communities.

Fig▶ Pine trees were planted as a windbreak near villages, which also contributed to Korea's traditional landscape.

Pine Trees and Forests Represent the Korean Identity

Pine is a species that has continually adapted to Korea's natural environment over the millinnea. Pine is considered to be the national tree of Korea because this tree has continued as the most beloved tree species in the country. The Korean landscape created by pine trees results from a long materialistic and spiritual relationship established between our ancestors and this tree. Thus, pine trees and pine forests are not only economic resources but also cultural resources that shape the Korean identity. Pine trees' role to reflect and improve Korea still continues. Usually, an identity of a country is established and continued by contemporariness and popularity. Although the pine tree was so closely intermingled with our ancestor's life, pine trees cannot remain a current symbol of Korean identity unless our modern lifestyle maintains a similar relationship. Such long-term, well-established identity cannot be supported by mere contemporariness.

In present-day Korea, pine trees are no longer used for farming implements, firewood, shipbuilding materials, or famine food. Nevertheless, pine trees continue its role as a national and cultural symbol even in the present industrialized society. Now, pine trees are planted around monumental buildings, in city green spaces, and along historic streets. Even today, pine trees continue to play an active role in shaping the cultural landscape of our country. A great example of pine trees' current role in Korea is the pine trees planted in the front yard of *Cheongwadae*, the presidential office and residence in Seoul.

Fig▸ The pine trees planted in the front yard of *Cheongwadae* (the Blue-tiled House), the presidential office in Seoul.

Fig▸ Windless Cold Pine, the pine forest at the entrance of *Tongdosa* Temple.

Traditional Cultural Landscape of *Tongdosa* Temple's Historic Pine Forest

8

It does not do *Tongdosa* temple justice to merely define it as the "Dharma Jewel Temple," one of the Jewel Temples of Korea. Its significance in Korea as the temple that keeps the relics and robe of Buddha cannot be overemphasized. However, what leaves the strongest impression on most visitors is the majestic pine forest that comes into view once they cross the *Mupunggyo* (Windless Bridge). Widely known as *Mupung-hansong* (Windless Cold pine), this quiet and secluded forest is praised as the foremost amongst the eight most beautiful spots of *Tongdosa*. Therefore, this pine forest is aptly named "the other jewel" of the Dharma jewel Temple *Tongdosa*.

As I walk in the midst of Windless Cold Pine, my entire body is enveloped by the sensory perceptions of pine. With each step comes the feel of pine, the colors of pine, the lighting of pine, and the scents of pine. A gentle breeze swirls around me. It is a special breeze, a pine breeze that is shaped by the special characteristics of a pine forest. As my walk continues, it seems as though I am walking in another world where pine colors, lighting, scents, and breezes come and go to tickle my perceptions. I walk leisurely, and a pristine joy passes through my body, and it seems to cleanse my soul.

Only the walkers can be the blessed recipients of this shiver of ecstasy.

The pine forest does not reveal its true beauty to those who speed through it in motorized vehicles. The sensibilities to appreciate nature do not come instantly; rather these sensibilities are developed over time and require some composure. As these sensibilities develop, one begins to notice the sounds of birds, wind, and streams. In addition, the "inner eye" helps to penetrate the silence that fills the space between sounds.

Therefore, such a walk in the pine forest is not a mere hike. Rather, it is a rite of passage through which one enters the "Pure Land", having purified accumulated karma. A walk in a pine forest is also a solemn ritual that initiates one to the sphere of stillness, compassion, and peace.

Monastic Pine Forest Represents Korean Sensibilities

Ancient temples are steeped in history, keeping traditional culture alive. Temples have not only preserved a myriad cultural assets for over a thousand years, but more importantly, they have carved out a uniquely Korean natural heritage that exists in harmony with the surrounding environment.

Currently, our accelerated industrialization is changing the topography of Korea every day; however, temples are the remaining oases that have maintained traditional natural scenery for several centuries. The pine forests, which have consistently stood by these temples, have become established as an iconic fixture of the traditional landscape in Korean minds, and they occupy an important corner of our hearts.

Constant human care is required for pine forests to thrive in Korea. To ensure the growth and health of pine forests for future generations, the forest floor must be cleared of dead pine needles, and hardwoods should be removed within the vicinity. Pine trees have their own developmental characteristics: being fond of sunlight, they cannot survive under the shade of hardwoods; their seeds cannot germinate in the composted dead

Fig▶ The Vulture Peak Hall (General Treasure No. 1471) and a three-story stone pagoda at *Tongdosa* temple.

Fig▶ Stupa garden containing the relics and monuments of the great master monks.

leaves.

In the agrarian society devoid of fossil fuel, dead pine needles were a good source of fuel that were traditionally collected as heating fuel for temples. The pine forest of *Tongdosa* temple was continually sustained because of constant intervention by humans. However, the human intervention in this pine forest ceased during the rapid industrialization of the past 40 years. As a consequence, nature regained control of ecological succession processes, thereby allowing hardwood trees, such as oaks, to exclude pine trees.

In *Tongdosa*, unlike most other places, these dwindling pine trees were conserved thanks to the foresight of its abbot, *Jeongwoo Sunim*. In his comment about the forest, we catch a glimpse of his painstaking efforts: "Mt. *Yeongchuksan* (*Vulture Peak*), in which *Tongdosa* is nestled, has a dense pine forest that has been cherished by many great masters. However, thicket growth prevented pine trees from developing, and

Fig▶ The Wisdom Dragon Ship painted on the outer wall of the Amitabha Hall at *Tongdosa* Buddhist temple.

in extreme cases caused them to die. Initially there was formidable objection, but with sincere efforts of communication I was able to convince the community to support a program to preserve the pine forest."

The Pine Forest that Envelopes the Stupa Garden

The eyes of this dendrology student have found another wonderful pine forest, other than the one near the entrance. This is the pine forest enveloping the stupa garden, a grove that was also reinvigorated by the recent removal of hardwood trees. The stupas and monuments of the great ancestral masters, which were once scattered around the temple compound, are now collected within this pine forest. Here, one can encounter such great masters as *Gyeongbong Sunim* and *Guha Sunim*,

Fig▶ The bell-shaped funerary stone stupa on the Diamond Ordination Platform and the pine tree.

who protected the pine forest of *Tongdosa* from Japanese plundering and supplied funds to the Korean interim government during the Japanese colonial period. This pine forest lends itself as a place for conversation between the present and past generations.

Another wonderful pine forest can be found on the mountain slope by the Diamond Ordination Platform. The Platform is the quintessence of the Dharma Jewel Temple *Tongdosa*, but it is a single, indigenous pine tree with red skin and bark that gives life to this essence. Most visitors barely notice this red pine that guards the stone bell-shaped funerary stupa seated in the center of the Ordination Platform, but I cannot help contemplating why it exists in this particular place. It is Korean tradition that the Mountain King Hall enshrines the Mountain King, with a tiger and a pine tree serving as temple guardians. The red pine tree that keeps guard of the Diamond Ordination Platform must be a sacred tree, which carries out the duty of protecting the Buddha's relics.

Chapter 1-1

Chun, Y. W. and Tak, K. 2006. The role of pine forests for shaping Korean traditional cultural landscape. In Cultural Heritage and Sustainable Forest Management: The Role of Traditional Knowledge. Proceedings of conference 8-11 June, 2006, Florence, Italy. Vol. 1: 177-182

Chapter 1-2

Chun, Y. W. 1999. Pine Tree. Forests and Korean Culture. Sumoon Publisher, Seoul. (in Korean)

Chapter 2

Chun, Y. W. 2006. Pine. Crossroads No. 5. The Asia Pacific Center for Theoretical Physics (APCTP) http://crossroads.apctp.org/

Chpter 3

Chun, Y. W. 1999. Pine Forests, the Face of the Land of Korea Altered by the Agrarian Culture. Forests and Korean Culture. Sumoon Publisher, Seoul. (in Korean)

Chapter 4

Chun, Y. W. 2001. Discovering Korea. Pine Forest of *Uljin*'s *Sogawang-ri* Village. Preserving Original Korean Pine. Koreana 15(3). The Korea Foundation.

Chapter 5

Chun, Y. W. 2009. Diamond Pine Forest at *Shingyesa*, Mt. *Geumgang*, North Korea. Buddhism and Culture. July 2009. No. 2. 18-22. Korea Buddhism Promotion Foundation.

Chapter 6

Chun, Y. W. 1999. *Kyungbok* palace, the main *Joseon* palace, under renovation with Korean red pine. Forests and Korean Culture. Sumoon Publisher, Seoul. (in Korean)

Chapter 7

Chun, Y. W. and Tak, K. 2006. The role of pine forests for shaping Korean traditional cultural landscape. In Cultural Heritage and Sustainable Forest Management: The Role of Traditional Knowledge. Proceedings of conference 8-11 June, 2006, Florence, Italy. Vol. 1: 177-182

Chapter 8

Chun, Y. W. 2009. Traditional Cultural Landscape of *Tongdosa* Temple's Historic Pine Forest. Buddhism and Culture. April 2009. No. 1. 62-67. Korea Buddhism Promotion Foundation.

Chun, Y. W. and Tak, K. 2006. The role of pine forests for shaping Korean traditional cultural landscape. In Cultural Heritage and Sustainable Forest Management: The Role of Traditional Knowledge. Proceedings of conference 8-11 June, 2006, Florence, Italy. Vol. 1: 177-182

Chun, Y. W. and Tak, K. 2008. Korea's traditional forest knowledge of managing fuel wood forests for Chosun white porcelain. Proceedings of The 1st International Conference on Forest Related Traditional Knowledge and Culture in Asia. 32-34

Chun, Y. W. and Tak, K. 2009. Songgye, a traditional knowledge system for sustainable forest management in Choson Dynasty of Korea. Forest Ecology and Management 257 (209) 2022-2026.

Chun, Y. W. A Treatise on the Forest Culture. Kookmin University Press, 1997. (in Korean)

Chun, Y. W. Pine and Our Korean Culture. Forests and Culture Society, 1993. (in Korean)

Chun, Y. W. Japanese cypress forest at Giso, Japan, managed for five hundred years. Hwangyongwoondong (Environmental Movement) June issue. Environmental Movement Yonhap 1997. (in Korean)

Chun, Y. W. 1999. Forests and Korean Culture. Sumoon Publisher, Seoul. (in Korean)

Chun, Y. W. 2004. Our Pine Tree We Should Know. Hyeonamsa, Seoul. (in Korean)

Chun, Y. W. 2005. Korea's Masterpiece Pine Trees. Sisailboneosa, Seoul. (in Korean)

Kim, B. S. 2001. New Edition of History of *Koryo* I. Sega I. Translated by Classic Studies. Shinseowon, Seoul.

Korea Forest Service. 2006. http://www.foa.go.kr/

Lee, Ki-baik. 1984. A New History of Korea. Translated into English by Edward W. Wagner. Ilchokak Publisher, Seoul.

Tak, K. Chun, Y. and Wood, P. 2007. The South Korea's Forest Dilemma. International Forestry Review. 9(1): 548-557